The Young Giants

A comparison of the early years of Newton and Einstein

by

Trevor Palmer

CONTENTS

Preface

This is a revised edition of *The Young Giants* first published as an ebook for Amazon's Kindle in February 2012. In the same year the Author published seven fiction ebooks one of which, *The Fifth Mind-leap*, touches on the early life of Albert Einstein – however, in a sci-fi context.

The Author, having moved in 2000 to the neighbouring village to that which saw the birth of England's greatest scientist, Isaac Newton, became a guide at the manor house of his birth in Woolsthorpe-by-Colsterworth. To have access to move freely about the rooms where the great man played out his early life, somehow combining his duties and responsibilities as the head of a sheep farm and Lord of the Manor with his religious and scientific interests, I found truly inspirational and awe-inspiring. Also, to see Albert Einstein's signature in the visitors book there provided the link which perhaps led to this book.

I had already tried to picture the childhood of farmer's son Isaac; his day-to-day routine as he struggled with the traumas of his early years and then, later, the trek to nearby Grantham for his preparatory schooling. Being a huge admirer of Einstein, I wondered how his early years compared.

By what seems to me an astonishing fact ... that both of these brilliant scientists had their most far-reaching insights in their early twenties ... a suitable conclusion could be reached in this comparison.

....

Acknowledgements:

Einstein ... *Einstein, The Life and Times* by Ronald W. Clark.

Newton ... *Let Newton Be* edited by John Fauvel.

Never at Rest by Richard S. Westfall.

Isaac Newton: Adventurer in Thought by A. Rupert Hall.

Volunteer staff and colleagues at the National Trust's Woolsthorpe Manor.

Introduction

Occasionally writers struggling to find a superlative for someone of really exceptional talent use the term *genius*. The term has been overused in recent years perhaps but, two men for whom this description was tailor made were Sir Isaac Newton and Albert Einstein. In this comparison of the two I have used the term *giant* and this merits an explanation.

It was modestly stated by Newton, not all that modest a man by nature, in the following pronouncement: he said, "If I have seen further than most men it is because I have stood on the shoulders of giants." A beautiful phrase I am sure you will agree and it refers, of course, to the use he made of the scientific advances or observations that other great men had already made and that he had made use of and built upon. Einstein was to repeat this same sentiment. It actually predates even Newton for Bernard of Chartres said something very similar in the early 12th century. But, whatever the modesty, our two subjects became themselves mighty giants for others to climb upon.

We are in awe of our geniuses: we idolise them. But to do so we need to have some sort of picture of them in our mind's eye. It is here that the media, sometimes unintentionally but often foolishly and frivolously can let us down. Both Newton and Einstein are usually presented as middle- to old-aged men whenever we see them and, indeed, looking at their faces, lined and carrying the wisdom of their years, they seem to fit the bill. The startling fact is, however, that *both* of these men produced their miraculous breakthroughs in the scientific world ***at an astonishingly early age!***

What follows is a comparative look at the early years leading up to their respective climactic discoveries; their school years, their later academic progress, and, finally, the magic of their moments of inspiration. This final revelation will, in itself, be quite a surprise to the non-scientists among the readers.

....

What the 'Giants' said and thought –
a few quotes

To those readers unfamiliar with Einstein … he showed a much more humorous side to his nature than did Newton. This first quote is typical ….

Einstein "Everything must be made as simple as possible. But not simpler."

Newton also had a view on simplicity ….

Newton "Truth is ever to be found in simplicity, and not in the multiplicity and confusion of things."

[**Author** *But were they right?*]

Here is how Newton saw himself ….

Newton "I do not know what I may appear to the world, but to myself I seem to have been only like a boy playing on the seashore, and diverting myself in now and then finding a smoother pebble or a prettier shell than ordinary, whilst the great ocean of truth lay all undiscovered before me."

Einstein saw Newton thus ….

Einstein "Nature to him was an open book, whose letters he could read without effort."

Newton's well-known statement about standing on the shoulders of giants was explored previously in the Introduction. Was the following by Einstein his own take on this? ....

Einstein "Every day I remind myself that my inner and outer life are based on the labours of other men, living and dead, and that I must exert myself in order to give in the same measure as I have received and am still receiving."

Three more of Einstein's witticisms ....

Einstein "Education is that which remains when one has forgotten everything learned in school."

"Common sense is nothing more than a deposit of prejudices laid down in the mind before you reach eighteen."

"Only two things are infinite, the universe and human stupidity, and I'm not sure about the former."

....

Beginnings

Two countries with strong cultural links, Germany and England, produced many of the world's finest philosopher-scientists from the seventeenth century onwards. It is perhaps fitting, therefore, that our two giants should come from these nations.

So, now we have the nationalities of our two. Let us look briefly at their family backgrounds and their births. Shall we find any indications there or in later years that either owed their genius to parents, to upbringing …. or to any other single cause?

Strangely, it was the city-born Einstein who was to achieve his successes on the outer fringes – some might argue *completely outside* – of the scientific world and of the sort of intellectually stimulating environment that is usually considered as conducive to producing original ideas of any kind. On the other hand, Newton, born in the tiny Lincolnshire hamlet of Woolsthorpe-by-Colsterworth, was surely destined – at least his mother saw it that way – to become a farm owner/manager. But this son of the land had a will of his own, as had the young Einstein, and this

led him to the more intellectually stimulating atmosphere of
Cambridge University.

....

Yes, Einstein was born in the industrial city of Ulm, which
lies on the river Danube about midway between Munich
and Stuttgart. The date was March 14th 1879. His parents
were non-practising Jews living a comfortable life-style.
His father, Herman, tried and ultimately failed to run a
small electrical and engineering workshop business. *He*
was certainly no genius, only his affable nature might be
seen in the later development of his son. The more affluent
side of the family was his mother's. Pauline nee Koch
brought into the marriage also, as did Newton's mother, a
little more cultured sophistication. Her son certainly
inherited one aspect of this, a love of music.

Only a year after Albert's birth the good-natured and
perhaps overly optimistic Herman saw his business
collapse and the family uprooted itself and left behind the
tallest cathedral spire ever built and the town whose people
were proud to say, "Ulmense sunt mathematici" [the people
of Ulm are mathematicians]. They took a small rented

house in the Bavarian capital, Munich, and Herman, this time in partnership with his brother, opened a small electrochemical works.

By coincidence it was at the University of Munich that one of Einstein's greatest scientific contemporaries, then in his early twenties, had taken up a post. This was the remarkable Max Planck. Their lives were to become intriguingly intertwined in later years.

When Albert was two years old, his sister Maja was born. Here then was set the scene and the characters for the early years of the life of the German giant. Imagine if you will those happy childhood days. Einstein himself later looked back fondly upon the regular Sunday excursions to the beautiful surrounding countryside, stopping at taverns for a beer for father and sausages for all. His relationship with younger sister Maja was, apparently, close and she became his confidante. Although both sides of the family could trace their Jewish ancestry back hundreds of years, they did not attend the synagogue or practise the taboos on what should be eaten. In fact, in the largely Catholic community of Munich, Albert's first school was one of that faith simply because *it was more convenient*.

What the young Albert did or thought about matters of science during these early years is unclear but a potential catalyst or milestone came when he was only five according to a famous anecdote. Being ill in bed his father showed him a pocket compass. It is said that the fact that the needle always pointed north no matter how the compass was turned planted the seeds in his young mind of the action of an invisible force. One aspect of his early school years *is* beyond dispute: he was extremely late in learning to speak! From this the Dyslexic Society deduced that this impairment may have been the reason and added his name to the illustrious number who also suffered this setback. The signs of dormant genius, though, were conspicuous by their absence. A concluding anecdote on this to ram the point home states that his father asked Albert's headmaster what profession his son should adopt. The all-knowing headmaster replied: "It doesn't matter; he'll never make a success of anything." At the age of ten, after five years at his primary school, Albert made the transition to secondary education, the Luitpold Gymnasium [this latter name is the German equivalent of our (English) grammar school].

How does all this compare with Newton's start in life?

....

Newton was to become known as a *difficult, perhaps prickly and argumentative, character and a rather isolated figure*. Just as we look for signs of his latent genius in his early life we may at least discern the signs of his darker nature.

The young Newton's father was also called Isaac. He was a man of the land, a typical yeoman farmer of middle England, prosperous through the breeding of the much-valued herds of sheep in this part of the country. He was, however, illiterate, signing his name with a cross. Whatever ambitions drove him on or whatever fate had in store for him, his star suddenly shone brighter when he married Hannah Ayscough. As in the case of Einstein's parentage it was the female side of the marriage that introduced learning and culture. To put the marriage on a suitable footing an ambitious father (Sir Isaac Newton's grandfather) bought for the married pair the Manor House at Woolsthorpe and the title and duties of Lord of the Manor that went with it. None-the-less, the Newtons, though top of the tree of working folk were still just that …

working folk, not gentry. Their farm was simply a business as was that of Albert's father.

On Christmas Day of 1642 young Isaac was born. But this was an unusual 'gift' package. Firstly, the young child was very tiny and not expected to survive. Secondly, his father had not lived to see his son born (and perhaps to have a pride in his achievements that were to follow). Nevertheless, the fine stone manor house must have wrung with sounds of joy, both seasonal and congratulationary, from servants and friends and family. The many visitors to the manor house today can recapture something of a feel for the occasion with a little imagination. The fire roaring in the huge fireplace in the kitchen and the smells of a goose cooking. A thousand miles away for the moment from the political turmoil of those times where roundheads and cavaliers would soon vie for supremacy in the neighbouring fields and towns. But, on that auspicious day Charles was still king.

Although the young Isaac not only survived, of course, he lived a long and healthy life. But mentally he was to suffer probably his first and worst trauma when still only a toddler of three years …..his mother re-married to the vicar of a nearby parish and moved out leaving him with his grandmother Ayscough.

Perhaps slowly but surely, an only child living comfortably but in turbulent times, the young boy found his place in the farming world of south Lincolnshire. His granny may have been strict but she probably had that warmth of compassion that most such ladies bear for their grandchildren. Just as the later Albert and his family turned *away* from their religious background it is likely that grandmother Ayscough (and his mother, who no doubt visited regularly) saw to it that young Isaac worshipped regularly and kept the faith.

In the neighbouring village of Skillington, young Isaac had three aunts, all with children. It is most likely, then, that the young boy was a frequent visitor and one probable outcome was that it may have been one of these aunts who put forward to his grandmother the name of some learned person in the village. For it was here and in the next village of Stoke Rochford that the small boy started his schooling, learning the basics in a so-called dame school.

At the age of 10 Isaac was possibly subject to some further trauma in his young life: his step-father died and his mother returned to Woolsthorpe Manor bringing with her three children from this second marriage. It was with his half-sister, Hannah, that Isaac was to form what was to be maybe his closest lasting friendship.

The next leap forward for our two subjects was their transition to what we now call secondary education. Let us look at the comparisons in this vital part of their lives.

The teenage years

To the young German boy the transition to higher education at the gymnasium was not a happy time. He balked at the discipline and at the way teaching did nothing to encourage free thought or to challenge the accepted wisdoms of the day. All this was retrospective opinion by the much older Einstein, however. But, the younger Einstein's mind certainly became set on challenging *received wisdom* even in his mid teens. Surprisingly, at the time of this educational move young Albert Einstein was still somewhat backward, however.

The boy had two favourite uncles in his family circle, Uncle Jacob, who may have had some influence on his learning, and Uncle C(e)asar, no intellectual, who he became particularly fond of. The anecdote concerning the former which suggests a connection with Albert's mathematical learning is that he is reputed to have said, "Algebra is a merry science. We go hunting for a little animal whose name we do not know, so we call it x. When we bag our game we pounce on it and give it its right name."

Uncle Casar was a prosperous grain merchant who had returned from Russia at the time of the boy's last year at primary school. He gave his nephew a gift, a model steam engine, perhaps to commemorate the coming educational move. Later, at the age of 16, Albert Einstein asked this uncle for his comments on his first scientific paper .... a remarkable effort for a boy of that age, concerning a major scientific puzzle of that time, the relationship between electricity, magnetism and the ether. In this he suggested certain experimental methods to resolve some of the problems.

Whatever his feelings about the teaching environment at the school, equating it to the typical German militaristic character, the somewhat introspective young boy must have made great strides forward to have tackled the subject matter of his paper to Uncle Casar. Certainly his maths had improved considerably and, as this period drew to a conclusion, he was already grappling with the mysterious scientific phenomena which were to dominate his thinking in forthcoming years. Before these final years were reached, however, there was another family crisis, again due to the failure of his father's business. This entailed the family leaving for Milan and Italy in 1894 leaving Albert, still with three years to go before he could get the diploma which would guarantee him a university place, to board. However, the now rather cocky and self-willed young fifteen-year-old had different ideas – as had, apparently, his school. After six months he left the hated gymnasium and followed his family over the Alps. There is some doubt but it is quite likely that he was expelled!

All this presented great problems for our young giant. In Germany a gymnasium certificate was an essential entrance paper to a university. A parallel with Newton's life also now brought a bearing on what was to occur. His father put pressure on his son to 'forget philosophical nonsense' and to take up a more practical career – in electrical engineering, of course. To do this it was decided to send Albert to the fine Swiss Federal Polytechnic School over the border in Switzerland at Zurich. Albert needed no diploma for this but *he was required to pass an entrance exam.* After some investigation it was decided he was too young at 16 (entrance was usually at 18) and would require some preparatory schoolwork. Given the thought that this path led to a career that would suit his father but would be anathema to him, is it to be wondered at that the iron-willed

Albert would ill-prepare himself and then fail the entrance exam? What now?

Perhaps fortunately, the principal of the ETH (to give the college its equivalent German designation) recognised Einstein's mathematical talents - and perhaps made appropriate allowances for his awkward character too. The outcome was that he again prepared in a local school for his entrance examination and *this time he passed!*

The school at which he made this final preparation was in the picturesque Swiss town of Aarau. The teaching here was friendly and all-embracing in contrast with the rigid discipline of the gymnasium.

A look at Einstein's character at this time compares very unfavourably with his later gentle and benign popular image. He was without doubt arrogant and, even, impudent, not just with his peer group but with his teachers. One thing his moves to Italy and now to Switzerland *had* shown him, however, was that other nationalities held an appeal to him much more than did his fellow Germans. This was to have a sudden and startling result. At the age of only 16 he shocked his father by firmly stating that he was renouncing his German citizenship and, as if that wasn't enough, he would sever any formal links with the Jewish faith. Yielding to this determined stance, Herman wrote to the German authorities and, in January of 1896, the deed was done. Albert Einstein was stateless.

....

Reaching back almost two hundred and fifty years before this, our other subject, having reached his twelfth year, now climbed onto the second rung of his schooling ladder. This was to the Free Grammar School at Grantham. This already ancient school was approximately eight miles north of Woolsthorpe along the Great North Road. Clearly a daily journey was out of the question so young Isaac took lodgings with a Mrs Clark at an apothecary's shop. Bear in mind that even as late as Einstein's days, young boys of 12 in the labouring classes were starting work.

The young Isaac probably had no greater enthusiasm for the disciplines of his new school than did the young Albert. He is known to have been bullied and to have found it difficult to form friendships with his peer group. What sort of things did this strong-willed but isolated young boy get up to?

Unlike our other subject, Isaac was strongly religious and, in his later teens he compiled a document which throws a beacon of light upon his teen years. During a period of intense religious feeling he suffered remorse for supposed sins he had committed and he listed these as follows …"Making a mouse trap on Thy day (Sunday). Idle discourse on Thy day and at other times. Missing Chapel (at school). Having unclean thoughts, words and actions and dreams. Falling out with servants. Peevishness with my mother. Refusing to go to the close (field) at my mother's command. Punching my younger sister. Squirting water on Thy day. Robbing my mother's box of plums and sugar. Calling Dorothy Rose a jade. Denying a crossbow to my mother and grandmother though I knew of it. Peevishness at Mr Clarks for a piece of bread and butter. Eating an apple in church. Stealing cherry cobs from Edward Storer. Using a fellow student's towel to spare my own. Setting my heart on money, learning and pleasure more than Thee." And, a more frightening one …"Threatening my father and mother Smith to burn them and the house over them." He may have not got on with his step-father. One can detect his strong nature in the above list (there were many other 'sins' listed) but they indicate a pretty normal childhood for any boy, I would have thought!

In addition to this document, Isaac purchased at 16 a small notebook for two-and-a-half pence in which he listed all sorts of information such as 'a bait to catch fish', 'a salve for sores' and 'a water to clear sight', etc.

As well as a clever mind Isaac proved very good with his hands, both at drawing and at making models. The walls at Woolsthorpe Manor still bear evidence of his skill in the former art scratched into the plaster work.

So, the young English giant progressed. But, just as Einstein's progress stuttered during this period, so did Isaac's ... His mother withdrew him from school to concentrate on running the farm. However, wiser heads were to prevail. His headmaster, Stokes – perhaps allied with Isaac's Uncle William Ayscough, who had been to Cambridge and was the rector of a nearby parish – persuaded Hannah to let him return to the grammar school to make his final preparation for university. On 5th June

1661, a young man of 18-and-a-half, Isaac Newton was accepted into Trinity College at Cambridge.

....

At the ETH Einstein, in apparent contrast with the lonely path that Newton was to take, developed friendships. His young manhood saw him develop into the sort of powerful, magnetic male figure that was attractive to the opposite sex. His looks did not let him down. He had a thick crop of dark, wavy hair and sported a wide moustache which drooped down slightly at the sides of a firm, sensuous mouth. He had a rounded face with a short cleft chin. But above all these attributes was the gaze of his dark and penetrating eyes.

On the down side was a growing forgetfulness which would often see him forget his key. Also, a carelessness in dress. Perhaps his mind was already burying itself in the important things in life, to him the mysteries of nature.

One of the friendships he developed was with a female student four years his senior. She was Hungarian and of peasant origins. Her name was Mileva Maric.

Two young men ... and a division of paths

In August of 1900, Albert Einstein finally graduated. His mark was 4.91 out of 6.00 ... not quite that of a genius, one might think. All his class-mates except the unfortunate Mileva passed too. But the cocky young graduate was now to receive a severe jolt to his pride. The hoped-for teaching job at the ETH was not forthcoming and suddenly the young man was in rather desperate straits for the Koch family, who had kindly sent him a regular allowance, decided it was time he fended for himself and stopped the payments.

In a sudden flurry of writing off for jobs, citizenship now assumed more importance. The outcome was that, with some effort, he obtained Swiss citizenship – which, incidentally, placed on him the obligation to do three months National Service. The year was 1901. A temporary job helped him through this difficult period until his hopes of a permanent job were at last realised. To anyone not familiar with Einstein's life, it would seem that this position must be his much sought after teaching post, perhaps at some university.

However, the job offer, which he quickly accepted, was as a civil servant in the Patents Office at Berne. It wasn't even at the class two grade for which he had applied. In view of his lack of engineering skills and knowledge of reading technical drawings and specifications, it was only at grade three. The starting salary was a modest 3,500 francs per year. Yet he was to stay in this job for seven long years and while sorting through and recommending applications for a variety of patents by day he was to launch into the world of science a theory that would stand the regular denizens of that world on their heads. How could he possibly achieve this miracle?

Before we look at how and why this came to pass we must slip back into the life of our other subject, for it would be partly on the results of Newton's own achievements that Einstein's would be built and measured.

....

Newton probably took to his university life very well yet, initially, his circumstances may have been just as fraught as were Einstein's. For a start, although agreeing to her eldest son's departure and one can imagine her misgivings and, perhaps, annoyance, Hannah Newton did not provide the financial support which her wealth would have permitted.

Isaac began his days at Trinity College as a sizar, a sort of fag for the wealthier students. Many of these had commenced their university careers as young as fifteen. Isaac was already a grown man by the standards of those times. All this would not have sat well with a young man of his sensitive and introspective nature but he had an iron resolve to learn and in learning to find the answer to many of Nature's riddles. This resolve was surely one he shared with the later genius of our tale. But, let us consider for a moment the times in which he lived ….

The mid 1600s was a time of strong religious beliefs … but, also, strong beliefs in the supernatural, the forces of the dark. Where a phenomenon could not be explained by logic, it was assigned to either divine or unknown dark forces. The new breed of observer and experimental scientist was only slowly and warily emerging. These scientist/philosophers had to be careful not to cross the teachings of the Church.

Let us look briefly at some of the unknowns.

Another remarkable contemporary of Newton's was Robert Hook. This slightly older man was into *everything*. With some squeamishness he dissected a living dog to try to find out just what function the lungs served, for it was not known then that a vital constituent of air was oxygen and that animals had to extract this and feed it into their bloodstream. He was also pushing forward the accuracy of chronometers, barometers and weighing machines. He found time to speculate about the nature of light and it was on this particular theme that he was to clash unpleasantly with Newton. There was so much to discover and scattered around the country was an army of amateur scientists trying to do just that.

The answers they sought might lie anywhere and to a man like Newton it was worth exploring any possible path to discover these truths about Nature.

One way he tried was through alchemy. At the college, Isaac Newton used the roaring fires of a laboratory he built, spending months in mixing metals and other substances, hoping to reveal … something.

He also looked for truths which he believed may have been discovered by the ancients and which might again be revealed if one studied the old biblical texts. One outcome, maybe, of this was that he came to dispute the idea of a Holy Trinity. His scientific mind could not accept that an almighty God could somehow share His power and he also claimed to have discovered proof that the Catholic Church had fraudulently altered documents to maintain this doctrine. This was to remain his dark secret, for revealing his thoughts would have put him into direct conflict with the established Church and Isaac was to become a very ambitious political man. To many people, perceiving Newton as, almost, the very embodiment of a scientist it would perhaps be surprising to learn just how religious he was. Nothing illustrates this more than the fact that a *quarter of a million words* written by him on this subject were auctioned at Sothebys long after his death. In *any* matter which seriously interested him, however, he went to extreme lengths to explore that subject and to then write prolifically about his conclusions. Newton would rate as one of the most thorough scientists of all time.

A continuing curse on the people of seventeenth century England was the so-called Black Death. Brought to the island's shores many centuries before Newton was born, this terrible killer plague waxed and waned over the years. A sudden outbreak at Cambridge in 1665 saw the colleges close their doors and send students and teachers home for two years. It must have seemed an annoying inconvenience for Isaac, now getting into his stride, to have to return to the idyllic or, perhaps, boring life at Woolsthorpe Manor.

Arriving back home for this long spell, Newton probably resolved not to let his scientific studies be side-tracked by farming affairs. We can picture this young man, maybe, after he had unloaded his belongings. He was not then the dignified man with the wig and flowing robes as depicted in the paintings and statues that his later years and his legacy would produce. He probably strode about energetically in a loose shirt with white stockings on strong young calves. His death mask now displayed at Woolsthorpe Manor shows a strong face with a large chin and a wide, down-turned mouth below a long nose. By repute, the mouth rarely turned in a smile. His gaze in his paintings shows a steady, haughty gaze. Then, at 22, he must have been quite a handsome young man and those eyes would have flashed with purpose.

But bored he did not intend to be and it is more than likely that the belongings he unpacked included many books together with instruments with which to experiment. These were placed in his study, a large room above the kitchen. One in ignorance might be forgiven for thinking that his major discoveries, to be enshrined in the later published work to be known in short as the 'Principia', would be revealed to him or by him over a period of several years and back in the more conducive environment of Trinity College. Astonishingly, this was not to be so. In this comparatively short period, while still at Woolsthorpe, Isaac Newton was to produce a breath-taking array of scientific discoveries, observations and laws which would propel his fellow men firmly into the scientific age. What was the gist of these?

First, he had "the theory of colours". Using prisms and sunlight shining through one of the study windows Newton split light into its colours and then re-constituted the white light. He also formulated his laws of motion and invented calculus to work out the paths of elliptical orbits such as the planets take. Perhaps he gazed out of the study window at the orchard below or simply took a contemplative stroll among the apple trees for some fresh air. It was at such a moment that his brilliant mind, now in overdrive, linked up the fall of an apple with the circling planets and derived from this his theory of a gravitational force. But, whatever the achievement in the theory it was a gigantic further step to then calculate the exact motions of those planets plus, later, visiting comets. In all its splendour this total effort was labelled his *anni mirabulis*. He himself put it another way, thus …

"I was in the prime of my age for invention and minded mathematics and philosophy more than at any time since."

Wow!

But, Newton was not the cocky young man that our other subject was to become at a comparable age. Sensitive as always to criticism and challenge, Isaac Newton was to keep his writings to himself for many years so, all this would not burst upon the world for many more years until his *Philosophiae Naturalis Principia Mathematica* and the later *Opticks* were published.

....

It was to be exactly 230 years later that our second subject was to stake his claim as a giant. The years leading up to this are very different from those lived through by Isaac Newton.

We have seen that the young Einstein, now a Swiss citizen, has at last got himself a job, be it one that sits oddly with his ambitions to teach and to discover the underlying forces of nature and the way they operate.

He himself was to say, "I want to know how God created this world. I am not interested in this or that phenomenon, in the spectrum of this or that element. I want to know his thoughts." Perhaps only the arrogance of youth could deliver such a statement but, he was to come as near to achieving this as any other man or woman.

The days leading up to his triumph were inauspicious. He was a much more social animal than Newton and, starting with a young private student to whom he taught physics to supplement his income, he formed a small group of friends around him that would discuss with him some of his ideas. But they were not quite the intellectual group that he might have had at a university despite the name they gave themselves, the Olympia Academy! It is more than likely that they fitted in a fair amount of drinking and general merriment as would any other group of young people.

During the years from 1900 to 1905 Einstein produced five minor dissertations concerning molecular forces. But his thinking was not *altogether* about matters of science.

In 1903 he married Mileva. Two of his Academy friends witnessed the quiet wedding and there was no honeymoon. Whatever his motives he was to settle down in a new apartment in Berne and, by the end of that year, his son Hans Albert was born. Einstein seemed able to compartmentalise his mind, simultaneously scribbling notes while dangling his young son on his knee. The setting for an earth-shattering miracle seems almost as remote as that of Newton unpacking his bags amid the cluttering hens at Woolsthorpe. Einstein was an unknown in the scientific world as well he might be. But, in 1905 he sent three papers for publication to the German scientific journal *Annalen der Physik.*

One of these tied in with his previous papers. It was to do with molecular forces and it concerned an oddity that had been observed for some time but which had defied explanation. The odd effect had been named after a Scottish scientist and was called Brownian Motion. If, for example, pollen dust is scattered onto water, its tiny particles zigzag about randomly and without any apparent decrease in their energy. This was a baffling phenomenon but our cocky young patents clerk, from some unknown backwater, not only showed in his paper how this could occur but along the way proved a fact that was hardly credited by most scientists of his day - that atoms and molecules truly exist. This was amazing stuff yet this was the least important paper of the three. What else did he come up with?

One of his other two papers was to be rated so important that several years later it would be awarded the coveted Nobel Prize. Ah, some might say; that would be his paper on Relativity. Amazingly that revolutionary and controversial paper which would turn accepted theory on its head was not the award-winning paper. The one that achieved that distinction was one laboriously entitled "On a

Heuristic Viewpoint Concerning the Production and Transformation of Light." What on earth was this all about, you may wonder?

Another unexplained riddle of science at this time was concerning a phenomenon to do with light. Let us briefly look at what was believed about the nature of this incredible form of radiation at the turn of the century. Newton had a favoured a view that light was made up of small particles of energy fired out almost like little bullets but even he was not too sure about this. Recent observations on light seemed to indicate strongly that it had a wave form like sound travelling through air or, simply, waves upon the sea. But this latter demanded a transmitting medium and for light this was thought to be a luminiferous ether, a so far undetected entity surrounding the Earth. Surely, whichever it turned out to be, light had to be in one camp or the other.

Other scientists were struggling to explain an observed phenomenon – that light striking metal receivers was strangely affected if a sheet of glass was interposed. There was not a convincing explanation from the greatest scientific minds of the day of why this should be so.

Einstein proposed that it was due to light being discharged in certain packets of energy. These were called quanta. He then showed that the reception of these would explain why electrons were ejected by the receiving metal and he further went on to calculate the kinetic energy of the electrons which were ejected. The phenomenon had been given the name *photoelectric effect*. Einstein gave a complete explanation of the whole process.

So, in just one year our giant had emerged from nowhere to explain two weird puzzles and to win himself a Nobel Prize. This should have been enough to establish himself as an intellectual wonder but, more – far more – was to come in his third paper.

Newton had, at the beginning of his *Principia* defined both space and time. These would be the definitions that would meet with our every day, common sense perspective of reality. They were the underpinning of his extremely accurate laws of motion: of *everything*. There may have been some brilliant philosophers who were to doubt these but Einstein went one better. In his Special Theory of Relativity (not the name *he* gave to his third paper), Einstein proposed a revolutionary concept of space and time which linked the two together and made them dependant upon the speed of the travelling body associated with them. At the age of only 26, this giant was to make even top scientists boggle at what he was saying. Because it seemed an affront to common sense, many thought the difficult to grasp arguments must be totally wrong but Newton had had to overcome the same sort of scepticism.

Einstein was to become one of the first of a new breed of scientist (of which our own Stephen Hawking is a modern exponent). Instead of observing and experimenting and, only then, providing a logical explanation for the experimental results, this new breed was to use their powers of imagination plus some very sophisticated maths to produce a theory. The experiments would then *follow* to confirm the theory.

Some of the experiments to verify the new concepts of the Theory of Relativity were going to be very difficult to perform due to the accuracy required. One outcome upon which the Theory insisted was that light passing near to a powerful gravitational source such as a star would be bent. It wasn't until the next solar eclipse that this could be and was verified. A cocksure Einstein was in no doubt himself that it would be!

Other predictions were more bizarre but all have been confirmed by our incredible modern technology. As to this Theory, why *Special* you may wonder? This is because it was limited to objects moving at a constant velocity. But, Einstein knew instinctively that it could also be extended to, say, accelerating bodies. This would take him another ten long years of slogging work to finalise for, unlike Newton who was as creative with the tool of maths as he was with anything else, Einstein needed to be *shown* which complex branch of maths he should adopt to get the correct results.

But, his earlier 9,000 word dissertation on a special form of relativistic movement entitled "On the Electrodynamics of Moving Bodies" must rank alongside the "Principia" in its innovative brilliance.

……….

This astonishingly early flowering of two mighty intellects, one at only 22 and the other at 26, surely places them worthily as Giants of Science, though the paths they each took to the moments of insight were, as shown in the above, on the whole so very different.

THE END

www.ingramcontent.com/pod-product-compliance
Lightning Source LLC
Chambersburg PA
CBHW051825170526
45167CB00005B/2162